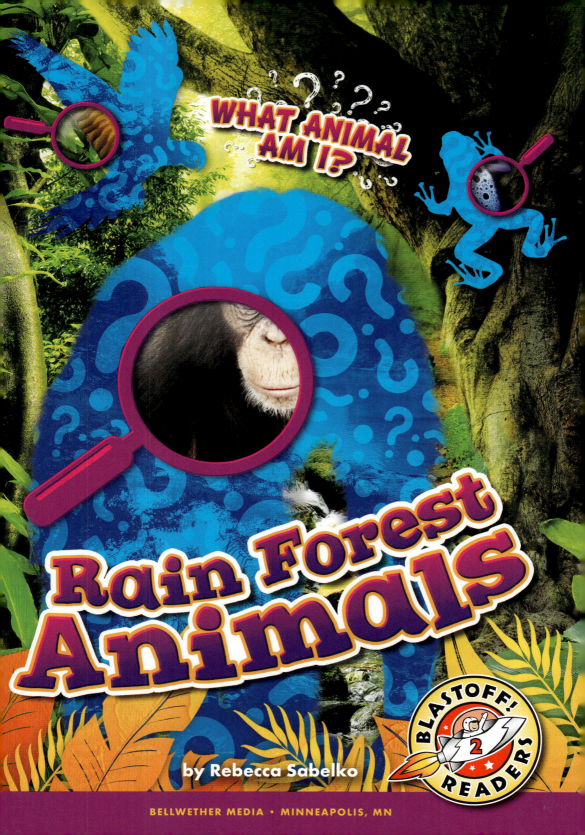

Blastoff! Readers are carefully developed by literacy experts to build reading stamina and move students toward fluency by combining standards-based content with developmentally appropriate text.

LEVELS

Level 1 provides the most support through repetition of high-frequency words, light text, predictable sentence patterns, and strong visual support.

Level 2 offers early readers a bit more challenge through varied sentences, increased text load, and text-supportive special features.

Level 3 advances early-fluent readers toward fluency through increased text load, less reliance on photos, advancing concepts, longer sentences, and more complex special features.

★ **Blastoff! Universe**

Reading Level

Grade K

Grades 1–3

Grade 4

This edition first published in 2023 by Bellwether Media, Inc.

No part of this publication may be reproduced in whole or in part without written permission of the publisher. For information regarding permission, write to Bellwether Media, Inc., Attention: Permissions Department, 6012 Blue Circle Drive, Minnetonka, MN 55343.

Library of Congress Cataloging-in-Publication Data

Names: Sabelko, Rebecca, author.
Title: Rain forest animals / by Rebecca Sabelko.
Description: Minneapolis, MN : Bellwether Media, Inc. 2023. | Series: What animal am I? |
  Includes bibliographical references and index. | Audience: Ages 5-8 | Audience: Grades 2-3 |
  Summary: "Relevant images match informative text in this introduction to different rain forest animals.
  Intended for students in kindergarten through third grade"-- Provided by publisher.
Identifiers: LCCN 2022009392 (print) | LCCN 2022009393 (ebook) | ISBN 9781644877319 (library binding)
  | ISBN 9781648347771 (ebook)
Subjects: LCSH: Rain forest animals--Juvenile literature.
Classification: LCC QL112 .S233 2023 (print) | LCC QL112 (ebook) | DDC 591.734--dc23/eng/20220307
LC record available at https://lccn.loc.gov/2022009392
LC ebook record available at https://lccn.loc.gov/2022009393

Text copyright © 2023 by Bellwether Media, Inc. BLASTOFF! READERS and associated logos are trademarks and/or registered trademarks of Bellwether Media, Inc.

Editor: Rachael Barnes   Designer: Brittany McIntosh

Printed in the United States of America, North Mankato, MN.

Table of Contents

Welcome to the Rain Forest!   4
Swinging in the Trees   6
Flashy Feathers   10
Color Warning   14
A Toothy Swimmer   18
Glossary   22
To Learn More   23
Index   24

Welcome to the Rain Forest!

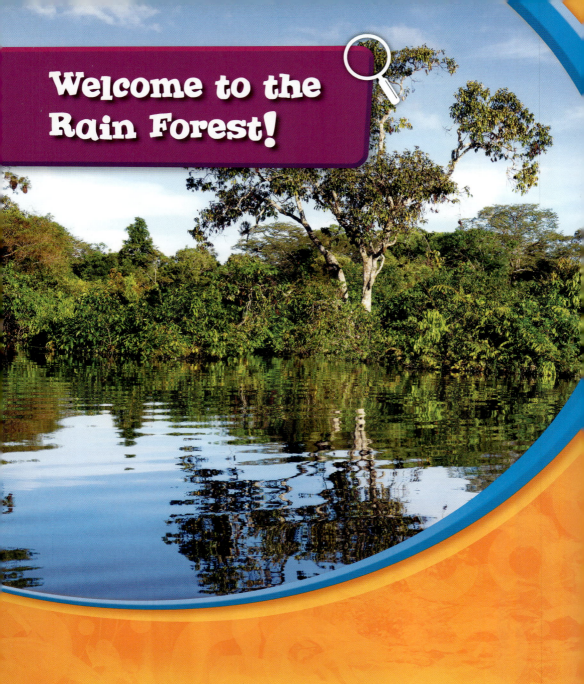

Rain forests are areas of tall trees that get a lot of rain.

They are home to over half of Earth's plant and animal **species**.

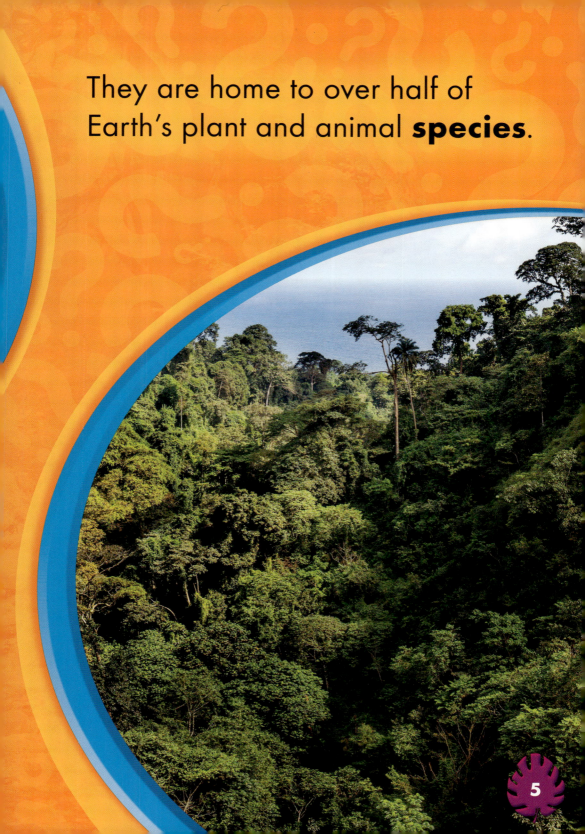

Swinging in the Trees

I am a **mammal** from Central and West Africa. I have brown or black hair.

My long arms help me swing between tree branches. What animal am I?

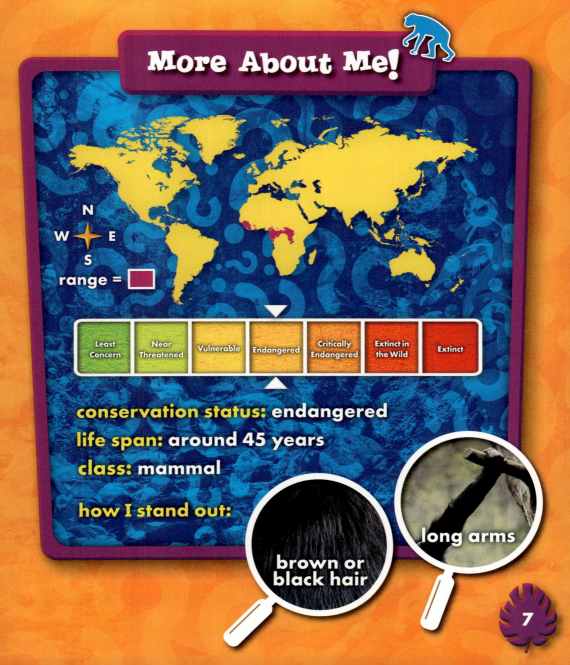

More About Me!

range =

Least Concern | Near Threatened | Vulnerable | Endangered | Critically Endangered | Extinct in the Wild | Extinct

conservation status: endangered
life span: around 45 years
class: mammal

how I stand out: brown or black hair, long arms

I am a chimpanzee! I live with my family.

I use sticks to dig for **insects** to eat. I use stones to smash open nuts.

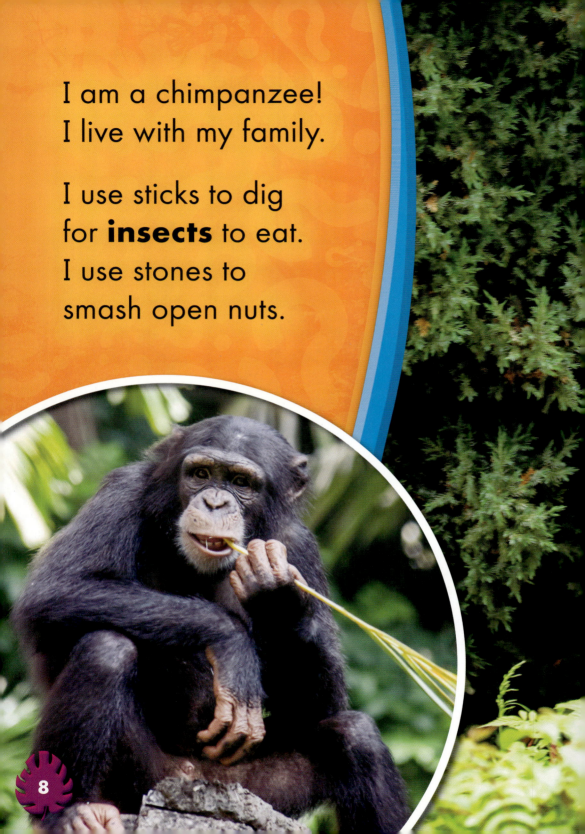

Chimpanzee Food

insects   nuts   fruit

Flashy Feathers

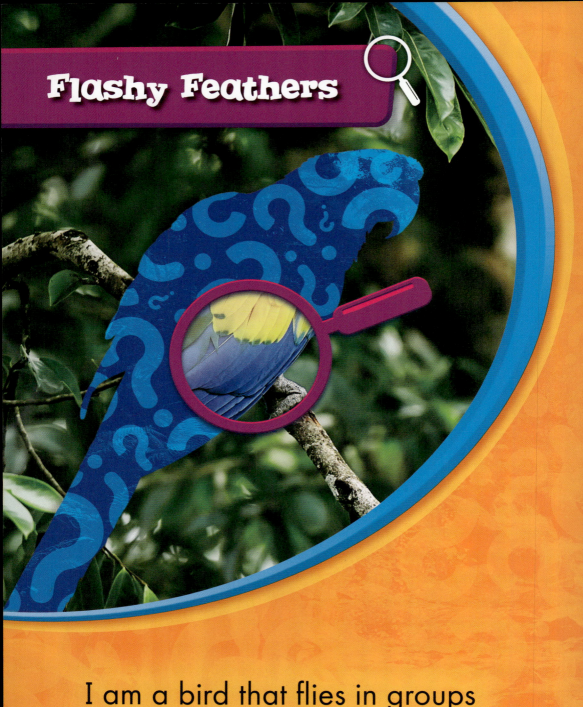

I am a bird that flies in groups through the rain forest.

My bright feathers stand out among tree leaves. What animal am I?

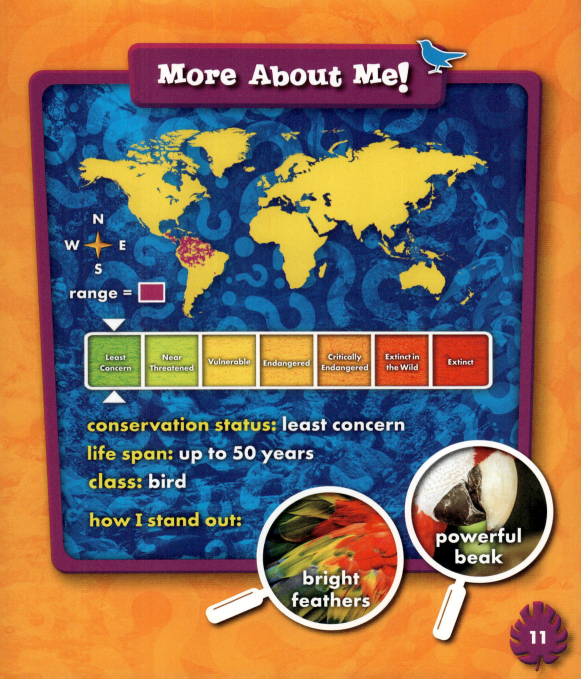

More About Me!

range =

Least Concern | Near Threatened | Vulnerable | Endangered | Critically Endangered | Extinct in the Wild | Extinct

conservation status: least concern

life span: up to 50 years

class: bird

how I stand out: bright feathers, powerful beak

I am a scarlet macaw! I squawk loudly to talk to other macaws.

My powerful beak cracks open nuts and fruit.

Color Warning

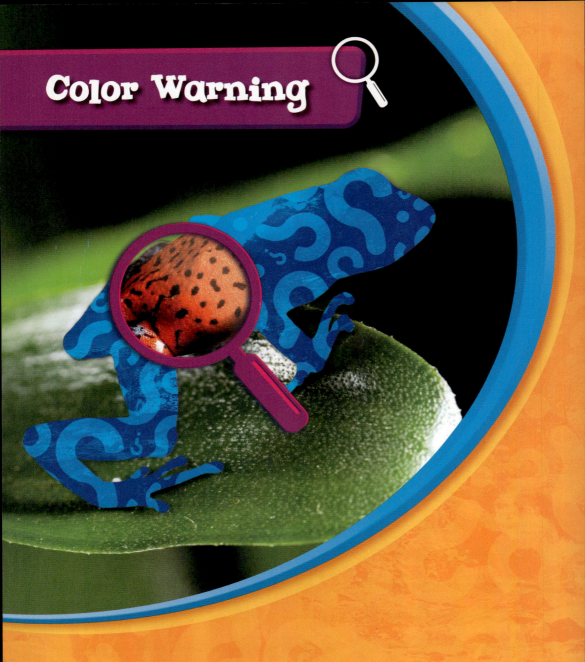

I am an **amphibian**. I live among leaves on the rain forest floor.

I am **poisonous**! My bright, colorful skin tells **predators** to stay away. What animal am I?

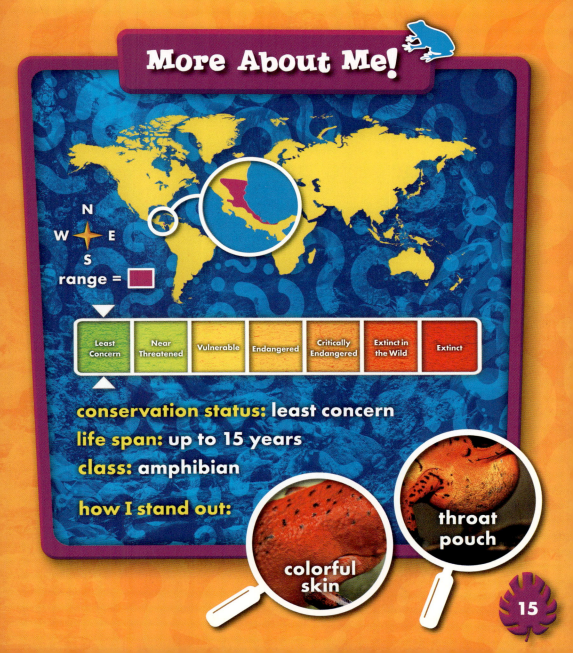

More About Me!

range =

Least Concern | Near Threatened | Vulnerable | Endangered | Critically Endangered | Extinct in the Wild | Extinct

conservation status: least concern
life span: up to 15 years
class: amphibian

how I stand out:

colorful skin

throat pouch

I am a strawberry poison dart frog! My throat pouch grows when I call to a **mate**.

I carry my **tadpoles** on my back to pools of water.

tadpole

Strawberry Poison Dart Frog Food

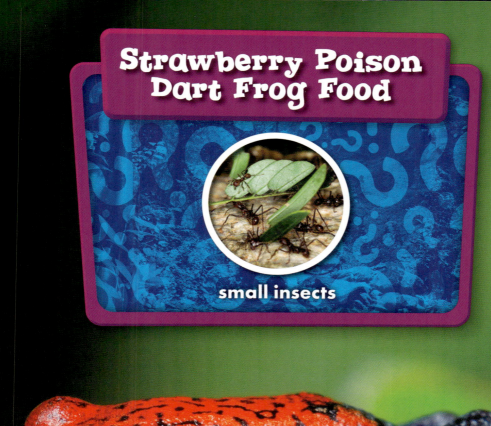

small insects

A Toothy Swimmer

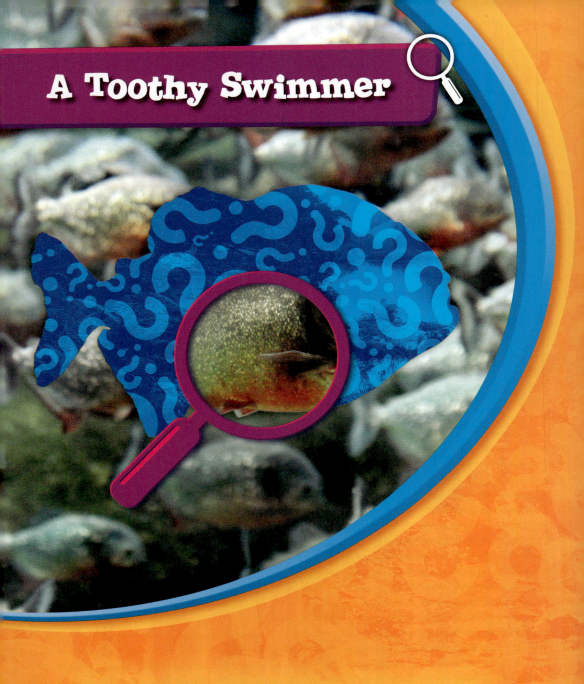

I am a fish. I live in rivers and lakes of the Amazon Rain Forest.

I have sharp teeth and strong jaws.
What animal am I?

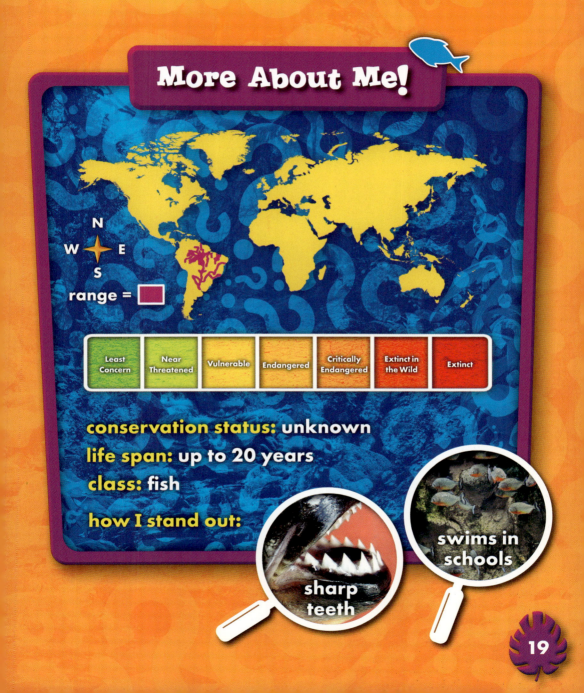

More About Me!

range =

Least Concern | Near Threatened | Vulnerable | Endangered | Critically Endangered | Extinct in the Wild | Extinct

conservation status: unknown

life span: up to 20 years

class: fish

how I stand out: sharp teeth, swims in schools

Red-bellied Piranha Food

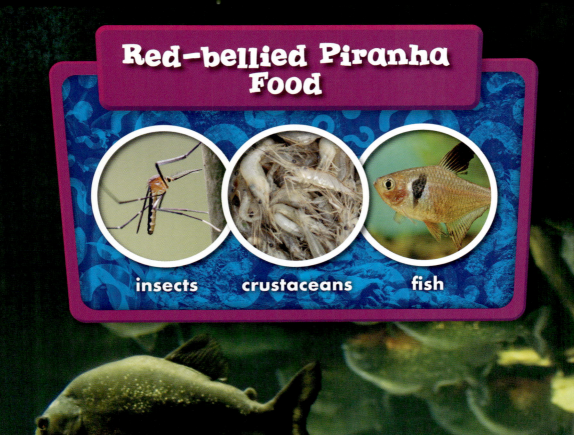

insects   crustaceans   fish

school

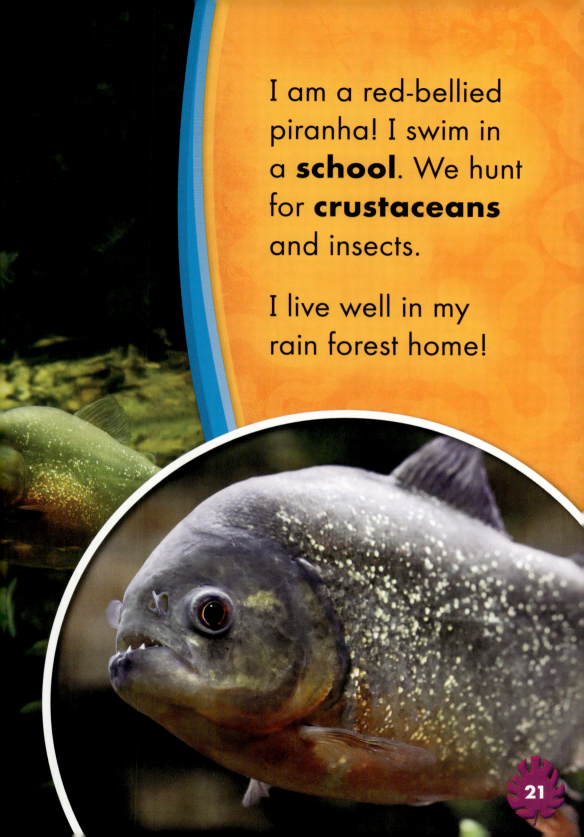

I am a red-bellied piranha! I swim in a **school**. We hunt for **crustaceans** and insects.

I live well in my rain forest home!

Glossary

amphibian—an animal that is able to live both on land and in water

crustaceans—animals that have several pairs of legs and hard outer shells; crabs and shrimp are types of crustaceans.

insects—small animals with six legs and hard outer bodies; an insect's body is divided into three parts.

mammal—a warm-blooded animal that has a backbone and feeds its young milk

mate—one of a pair of adult animals that produce offspring

poisonous—causing sickness or death when touched or eaten by animals

predators—animals that hunt other animals for food

school—a group of fish

species—kinds of animals

tadpoles—baby frogs

To Learn More

AT THE LIBRARY

Kenney, Karen Latchana. *Macaws*. Minneapolis, Minn.: Bellwether Media, 2021.

Labrecque, Ellen. *Day and Night in the Rain Forest*. North Mankato, Minn.: Pebble, 2022.

Statts, Leo. *Piranhas*. Minneapolis, Minn.: Abdo Zoom, 2019.

ON THE WEB

Factsurfer.com gives you a safe, fun way to find more information.

1. Go to www.factsurfer.com.

2. Enter "rain forest animals" into the search box and click 🔍.

3. Select your book cover to see a list of related content.

Index

Africa, 6
Amazon Rain Forest, 18
amphibian, 14
arms, 7
beak, 13
bird, 10
chimpanzee, 6–7, 8–9
colors, 6, 15
Earth, 5
family, 8
feathers, 11
fish, 18
food, 8, 9, 12, 13, 17, 20, 21
hair, 6
jaws, 19
lakes, 18
mammal, 6
mate, 16
more about me, 7, 11, 15, 19
poisonous, 15
predators, 15

rain, 4
red-bellied piranha, 18–19, 20–21
rivers, 18
scarlet macaw, 10–11, 12–13
school, 20, 21
skin, 15
species, 5
squawk, 13
strawberry poison dart frog, 14–15, 16–17
swim, 21
tadpoles, 16
teeth, 19
throat pouch, 16
trees, 4, 7, 11

The images in this book are reproduced through the courtesy of: Eric Isselee, front cover (chimpanzee); Passakorn Umpornmaha, front cover (parrot); Alex Stemmers, front cover (frog); Teo Tarras, front cover (background); chittakorn59, p. 3; Jess Kraft, p. 4; imageBROKER/ Alamy, p. 5; Abeselom Zerit, pp. 6, 7 (right); Marla_Sela, p. 7 (left); Danny Ye/ Alamy, p. 8; Helen Davies/ Alamy, pp. 8-9; ChaiyonS021, p. 9 (top left); Don Mammoser, p. 9 (top middle); K.Sek, p. 9 (top right); Petr Salinger, p. 10; Narupon Nimpaiboon, p. 11 (left); Maciej Czekajewski, p. 11 (right); Alvaro Trabazo Rivas, p. 12 (top left); DeawSS, p. 12 (top right); Martin Mecnarowski, pp. 12-13; Michele Lynn jasen, p. 13; Dirk Ercken, pp. 14, 15 (left); Jeroen Mikkers, p. 15 (right); rikrik, p. 16; George Grall/ Alamy, pp. 16-17; Kevin Wells Photography, p. 17 (top); Hayati Kayhan, p. 18; guentermanaus, p. 19 (left); Andrey Novgorodtsev, p. 19 (right); Salparadis, p. 20 (top left); Tarcisio Schnaider, p. 20 (top middle); Roberto Dani, p. 20 (top right); Amazon-Images/ Alamy, pp. 20-21; Grigorii Pisotsckii, p. 21; Jones M, p. 22; Dewin ID, p. 23.